A special thanks
to everyone
who has helped make
Know Yourself
what it is today.

Dear Reader

Knowing yourself is truly the beginning of all wisdom. We give young learners the building blocks they need to start their unique journey of self-discovery: an understanding of human anatomy — literally how we are put together. Knowledge of one's own human body is an empowering context on which anyone can build.

Learning about the body and mind at a young age sets the foundation for honoring one's physical form, develops self-esteem and self-confidence, and begins the discovery of who we are meant to be in this world.

Now that's real power.

The Know Yourself Team

ADVENTURE 5

Quick-Start Guide

Hello Know Yourselfers!

Follow these steps to start a new journey and explore the digestive system. Have fun on this Self Literacy quest and remember - Keep your wits about you!

1

Grab a kimono and fish! We're going to Japan!

Find Japan on your atlas, or find an online map of the world.

2

Read Time Skaters Adventure 5.

Pinky and Naz encounter Tokugawa Ieyasu and his newly formed shogunate. This time their problems are on a whole new scale.

3

Get equipped!

Gather your supplies and prepare for your activities. Every great adventure begins with the smallest parts.

III

Table of Contents

QUICK-START GUIDE III

    Hello Adventurer! 1

TIME SKATERS ADVENTURE 5 3

LEARNING CALENDAR 28

    Home Inventory Checklist 30

PART 1 • KNOW YOUR HISTORY 32

    Know Your Ninjas 36

ACTIVITIES:
- From Hokku to Haiku 38
- Fashion Statements 42
- Time For Tea 44
- Neat-o Edo Art 46
- Natural Talent 50
- Fold Your Words 52
- The Shogun Must Go On! 54

PART 2 • KNOW YOUR DIGESTIVE SYSTEM 56

Royal Flush 62

Gas: A Work O'Fart 64

ACTIVITIES:
- Digestion Detective 70
- Follow Your Food 74
- A Fuzzy Situation 78
- Numbers and Nutrients 80
- Whole-y Guacamole! 84
- Digestive Dash 86
- Digest Your Knowledge 88

PART 3 • KNOW YOUR APPETITE 90

RECIPES:
- Pickled Cucumbers 92
- Sushi Rolls - Vegetable Maki 94

Thoughts for Young Chefs 99

PART 4 • FOOD FOR THOUGHT IN FEUDAL JAPAN 102

Further Reading 107

Hello Adventurer!

Welcome to Adventure 5 - The Digestive System.

In this workbook, you will learn about Ancient Japan and your body's Digestive System. There will be information to read, activities to complete, and quizzes to take when you are ready to challenge yourself! Take your time along the way - spend as much or as little time as you like on each activity.

Good luck, and have fun!

Can you find Japan?

THE TIME TRAVEL CLOCK READS 1611

Get ready to digest some knowledge!

ADVENTURE 5

LEARN ABOUT
The Digestive System
Get the breakdown on how your body turns food into the nutrients it needs

VISIT
Edo Period Japan
While it is a time of peace, dangers lurk from those who wish for conflict

MEET
Tokugawa Ieyasu
one of the Great Unifiers of Japan

ようこそ
(Yokoso)*
That means "Welcome!" in Japanese.

*Say it like this: "yoh-ko-so"

Note: Syllables in Japanese words are generally spoken with equal stress, and for the same length of time.

Follow us to...

ADVENTURE 5

Time Skaters Adventure 5: Hard to Stomach

THE DIGESTIVE SYSTEM

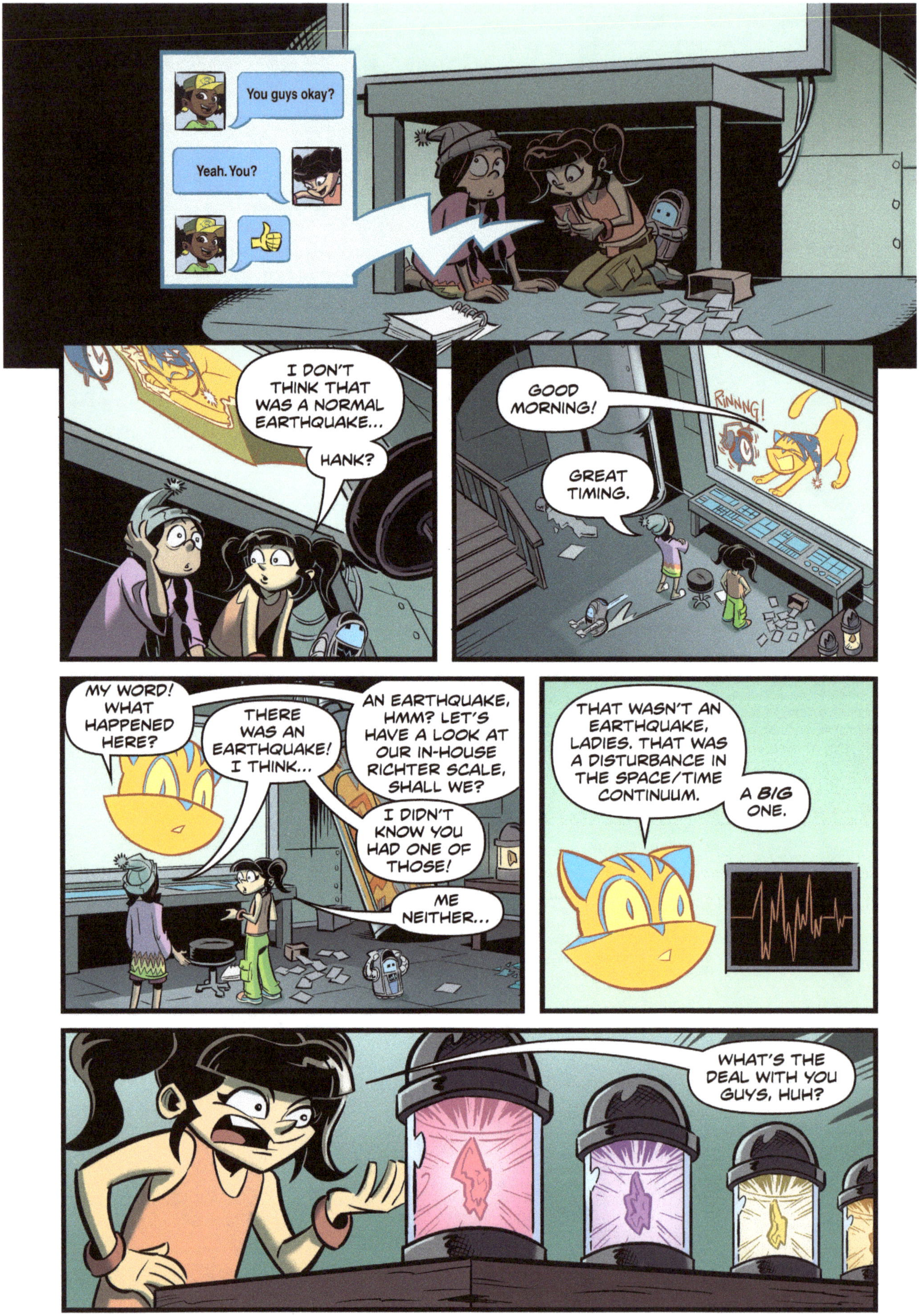

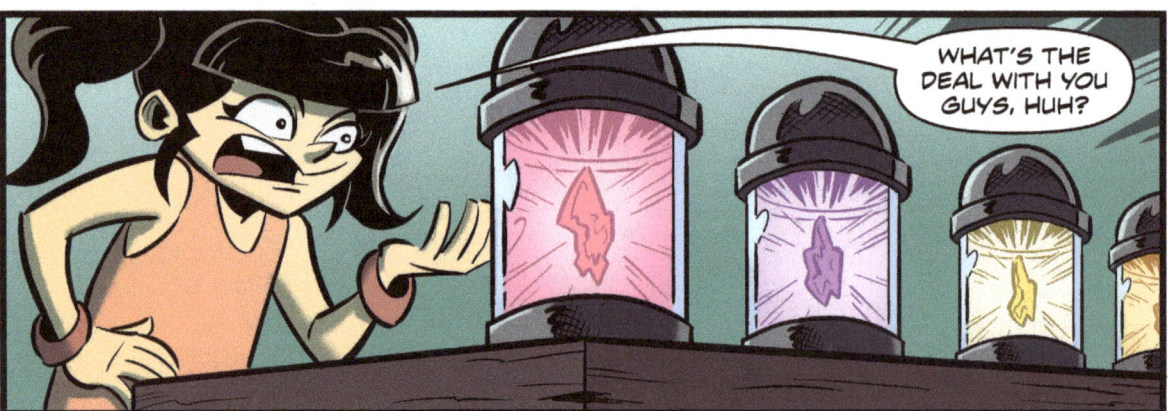

THE DIGESTIVE SYSTEM

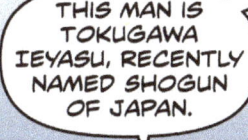

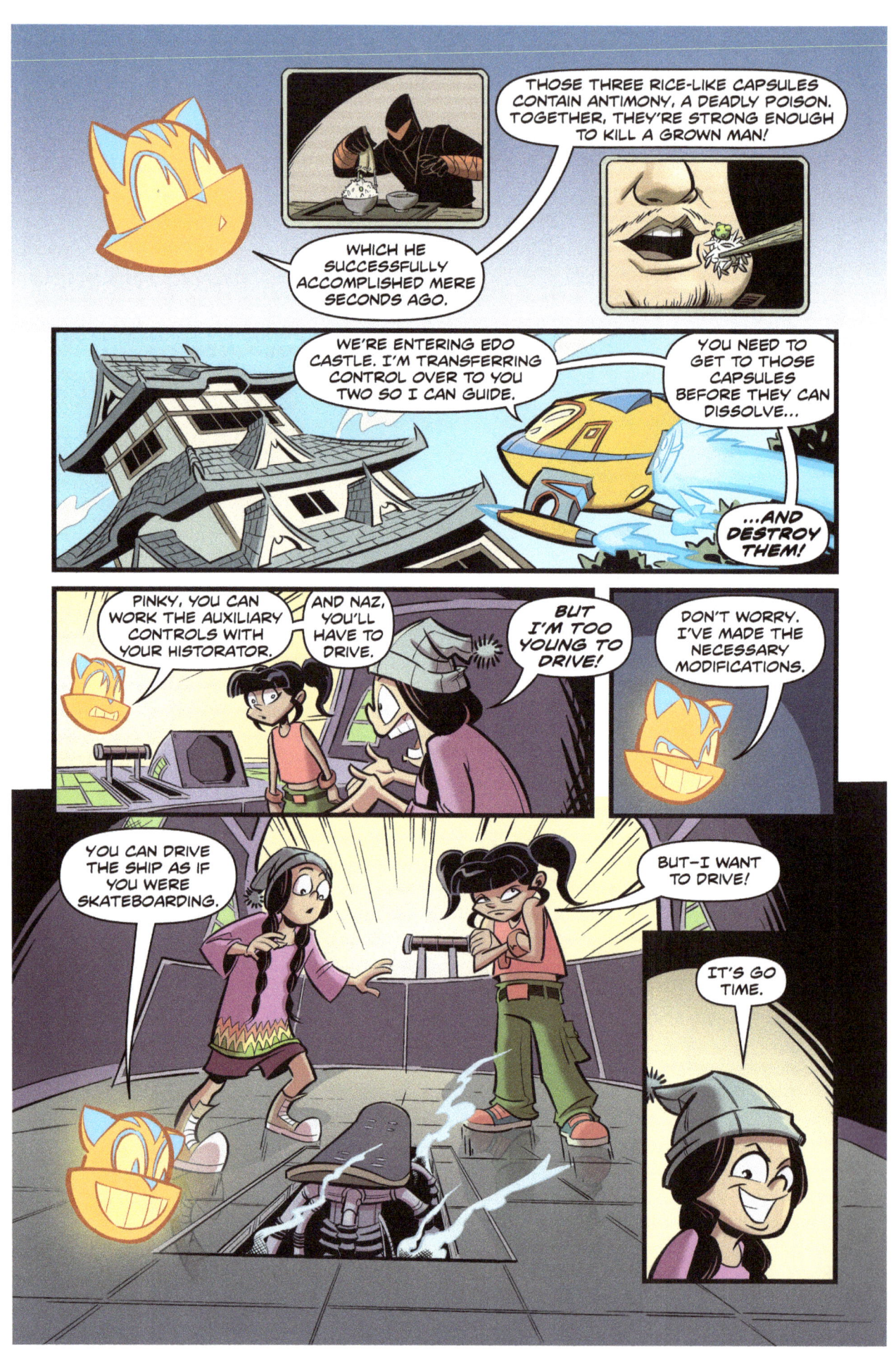

THE DIGESTIVE SYSTEM

*DID YOU KNOW...

THAT WHEN YOU SWALLOW FOOD, YOU USE MORE THAN 20 PAIRS OF MUSCLES TO SEND IT DOWN INTO YOUR STOMACH?

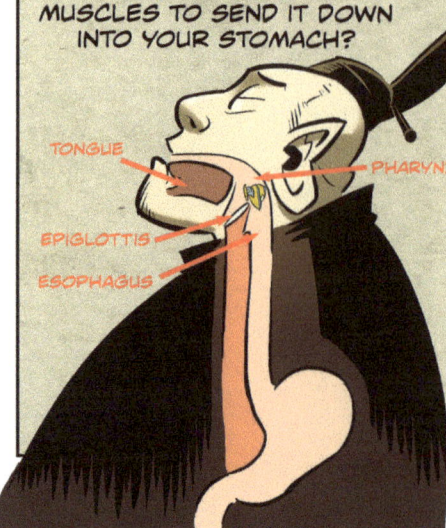

*THE TONGUE MUSCLES SEND THE FOOD INTO YOUR **PHARYNX** WHERE A FLAP CALLED YOUR **EPIGLOTTIS** COVERS YOUR WINDPIPE TO PREVENT FOOD FROM "GOING DOWN THE WRONG PIPE" AND TO PREVENT CHOKING.

*FOOD THEN ENTERS THE **ESOPHAGUS** – A 6 TO 8 INCH LONG MUSCULAR TUBE THAT LEADS TO YOUR STOMACH. THE MUSCLES IN YOUR ESOPHAGUS CONTRACT IN A SPECIAL WAVE-LIKE PATTERN CALLED **PERISTALSIS**.

*THE MUSCLES IN FRONT OF THE FOOD RELAX WHILE THE MUSCLES BEHIND IT CONTRACT TO PUSH IT FORWARD AND PREVENT IT FROM GOING BACKWARDS.

*SAY THEM LIKE THIS: | PHARYNX - "*FAIR-INKS*"
EPIGLOTTIS - "*EP-EH-GLOT-IS*"

ESOPHAGUS - "EH-SOF-EH-GUS"
PERISTALSIS - "PAIR-EH-STALL-SIS" | SPHINCTER - "SF-INK-TER"

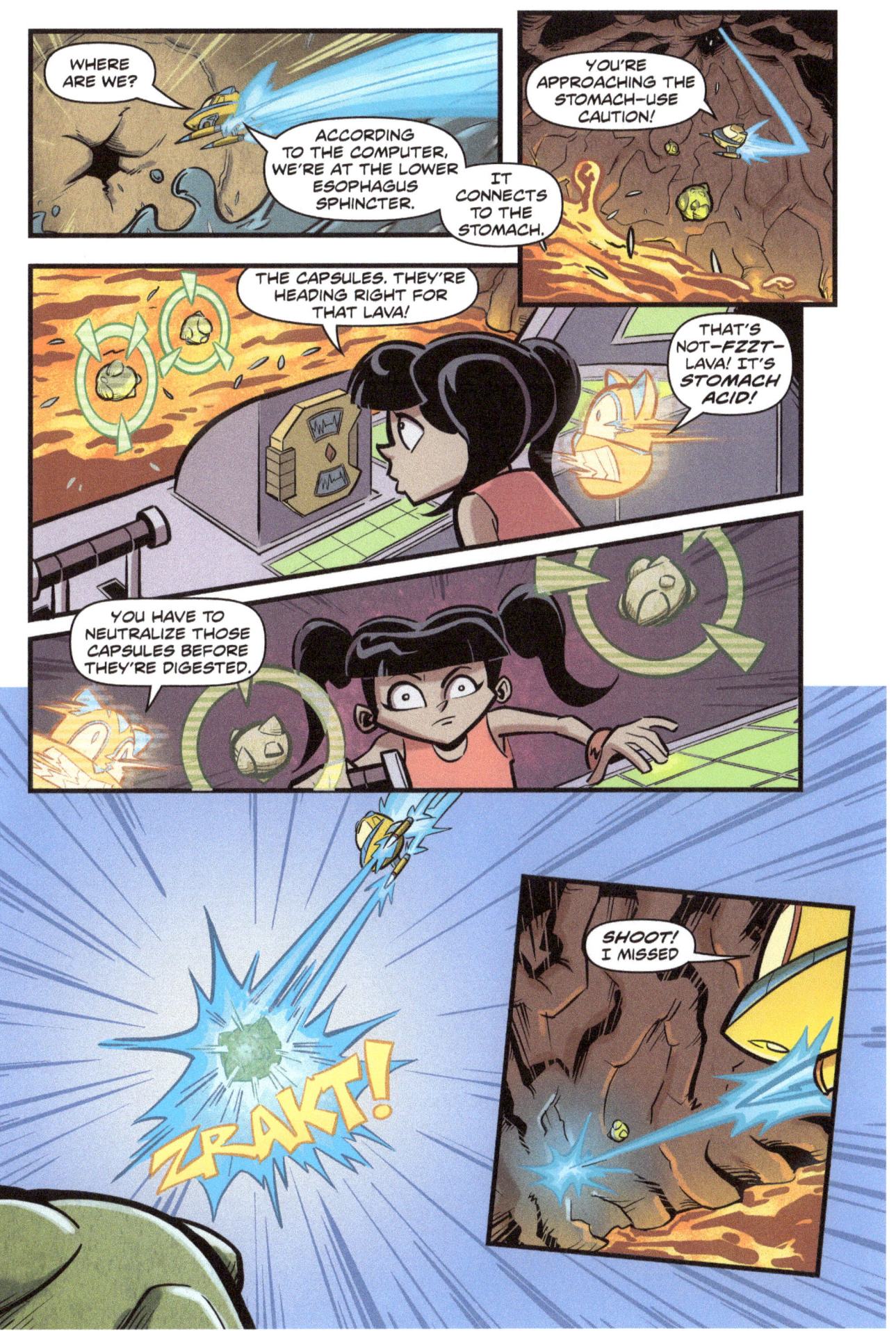

coloring opportunity

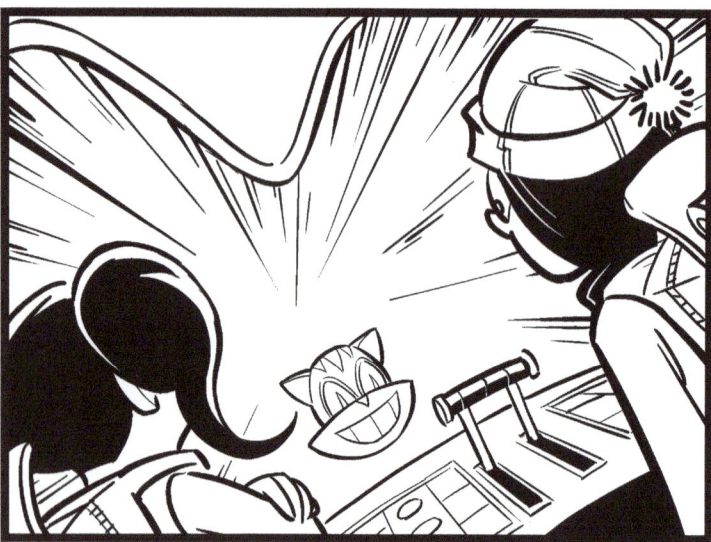

ADVENTURE 5

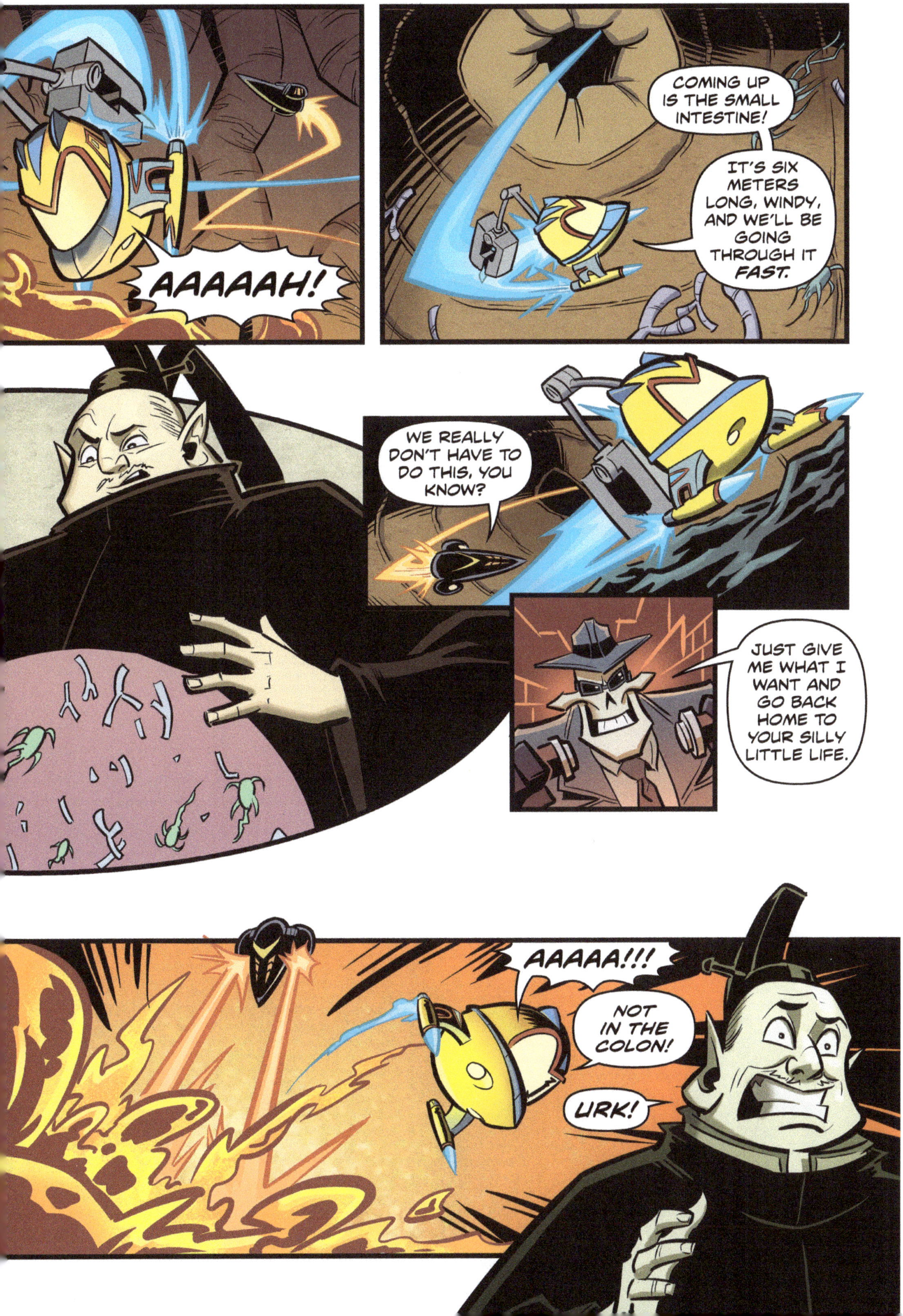

DON'T TOUCH IT.

NAZ?! ARE YOU OKAY?!

YEAH. I THINK SO...

BURP!

BEAUTIFUL. IT REALLY IS!

HOW DID YOU KNOW WHAT I WAS THINKING?

I...I DON'T KNOW.

...and the story continues!

27

Learning Calendar

Part 1

Know Your History

Estimated hours 5 hours of fun

Gather the adventure equipment you'll need from around the house - find the checklist on pages 30 and 31!

Locate Japan on the map on page 1.

Read the comic **Time Skaters Adventure 5 - Hard to Stomach**. Find it at the beginning of this Adventure Guide!

Get to *Know Your History*.

Uncover *Know Your Ninjas*.

Make Haste *From Hokku To Haiku*.

Find *Your Fashion Statements*.

Paint *Neat-o Edo Art*.

Fold *Your Words*.

Discover *The Shogun Must Go On!*

Part 2

Know Your Digestive System

Check out *Know Your Digestive System*.

Become a *Digestion Detective*.

Follow *Your Food*

Count *Numbers and Nutrients.*

Do the *Digestive Dash.*

Digest *Your Knowledge.*

Part **3**

Know Your Appetite

Explore *Know Your Appetite.*

See the recipes on the following pages. Make a shopping list, purchase ingredients, and get your kitchen ready!

Make *Pickled Cucumbers* and *Sushi Rolls.*

Share your dishes with your family. Discuss *Thoughts for Young Chefs* around the table.

Part **4**

Show What You Know!

Spot *Food for Thought in Feudal Japan.*

Check out *Further Reading* for more opportunities to learn.

Great job Adventurer!

Home Inventory Checklist

Ask your parents to help you find these items around the house. These are some of the tools you will need on your adventure.

- ☐ **Paper**
  - From Hokku to Haiku, Numbers and Nutrients
- ☐ **Pen or Pencil**
  - From Hokku to Haiku, Neat-o Edo Art, Numbers and Nutrients, A Fuzzy Situation
- ☐ **Two six-sided die**
  - Numbers and Nutrients
- ☐ **Pennies, beads or another token**
  - Numbers and Nutrients
- ☐ **Newspaper**
  - Neat-o Edo Art
- ☐ **Thick paper or canvas**
  - Neat-o Edo Art
- ☐ **A cup**
  - Neat-o Edo Art
- ☐ **Water**
  - Neat-o Edo Art, Follow Your Food
- ☐ **Paints**
  - Neat-o Edo Art
- ☐ **Paintbrushes**
  - Neat-o Edo Art
- ☐ **A few paper towel sheets**
  - Neat-o Edo Art

Be sure to check the items off when you've found them.

- [ ] **Posterboard**
    - Digestion Detective

- [ ] **Markers, craft paint, or other materials**
    - Digestion Detective

- [ ] **2 pieces of yarn or string, 6 meters and 3 meters long**
    - Digestion Detective

- [ ] **Liquid glue**
    - Digestion Detective

- [ ] **1 small resealable plastic sandwich bag**
    - Follow Your Food

- [ ] **Cereal**
    - Follow Your Food

- [ ] **Soda**
    - Follow Your Food

- [ ] **1 nylon stocking**
    - Follow Your Food

- [ ] **Scissors**
    - Follow Your Food, A Fuzzy Situation

- [ ] **A fuzzy sock (that you don't need)**
    - A Fuzzy Situation

- [ ] **Square-inch grid paper**
    - A Fuzzy Situation

Don't worry if you can't find every single item – just use your imagination to find substitutions!

THE DIGESTIVE SYSTEM

Know Your History

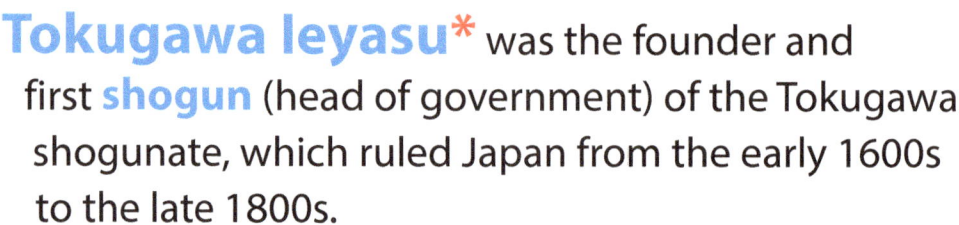

Tokugawa Ieyasu* was the founder and first **shogun** (head of government) of the Tokugawa shogunate, which ruled Japan from the early 1600s to the late 1800s.

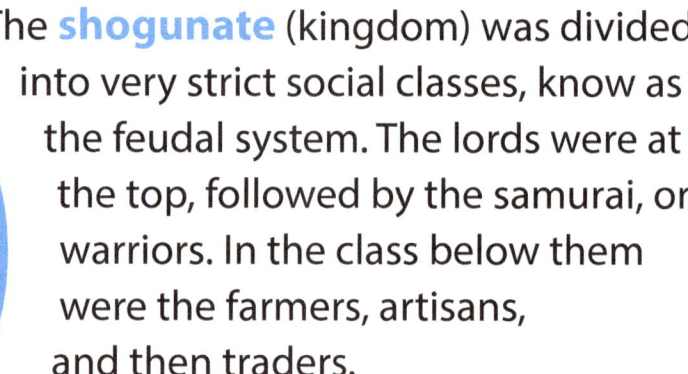

The **shogunate** (kingdom) was divided into very strict social classes, know as the feudal system. The lords were at the top, followed by the samurai, or warriors. In the class below them were the farmers, artisans, and then traders.

> ***Say it like this:**
>
> **Tokugawa Ieyasu**
> "**toe-koo-gah-wah ee-ye-yah-soo**"
>
> Note: Syllables in Japanese words are generally spoken with equal stress, and for the same length of time.

ADVENTURE 5

Edo Period

The Edo Period was a time of great cultural and economic growth in Japan. Many of the roads and water transportation systems developed back then still survive today.

More than roads and plumbing, the Edo Period was a time when Japanese culture was transforming.

Most of the art, food, and traditions of modern-day Japan, like sushi rolls and fashion, became popular in the Edo Period.

Know Your History

Tokyo

Edo was the original name for the modern city of Tokyo, and it was the seat of power for the Tokugawa shogunate. The city was a strategic port for shipping, and it has always been famous for its bustling fish markets.

Today, Tokyo is the largest city, not just in Japan, but in the whole world!

coloring opportunity

THE DIGESTIVE SYSTEM

Know Your Ninjas

Shinobi,* the Japanese word for ninjas, were spies as well as assassins. When you think about ninjas, you probably imagine men in black masks with swords. But not all ninjas were men, and some of them hid in plain sight.

Kunoichi,* the female counterpart to ninjas, were just as deadly and well-trained, and even had some advantages over men. Most lords and samurai warriors had personal guards to protect them from attackers, so even the stealthiest ninja would have trouble sneaking into their well-guarded fortresses or castles.

On the other hand, many kunoichi worked undercover, disguised as maids, entertainers, or domestic workers.

In these disguises, they could walk into places that male ninjas couldn't break into and, once there, they could gather information without raising any suspicions.

Kunoichi had to be skilled in hand-to-hand combat, using weapons like daggers, poison, bladed fans, and slip-on metal claws called **neko-te**.* These were effective, as well as easy to hide in a pocket or sleeve.

*Say them like this:

Shinobi - "she-no-bee"
Kunoichi - "kuh-no-ee-chee"
Neko-te - "knee-koe-tay"

Note: Syllables in Japanese words are generally spoken with equal stress, and for the same length of time.

THE DIGESTIVE SYSTEM

From Hokku to Haiku

The **Edo Period** was a time of overall peace for much of Japan, which resulted in many changes to society. Samurai who had previously been needed for their military expertise found themselves with time to branch out and fill new roles. **Matsuo Munefusa** had put aside his interests in poetry to become a samurai, but when his feudal lord died in 1666, he took up the pen name Matsuo Bashō and devoted himself to his art.

The **haiku** (or **hokku**, as it was known at the time) had become a somewhat stale form of poetry, but Bashō drew on his Zen philosophy to turn it into something that would speak to all ages.

Centuries later, the haiku still draws the excitement of millions with its ability to combine simplicity and depth.

This is one of Bashō's earliest haiku

From Hokku to Haiku

Now make your own haiku! While Matsuo Bashō tended to write about nature, a haiku can be about any topic that excites you, as long as you follow the 5-7-5 syllable form. Syllables are the sound sections in a word — here are some examples:

Know (know) is **one** syllable

Yourself (your-self) is **two** syllables

Adventure (ad-ven-ture) is **three** syllables

Try counting the syllables in Matsuo Bashō's haiku on the previous page. Did you count 5 syllables in the first and last lines and 7 in the middle line? If so, you're ready to write your own!

Materials:

- **Pen**
- **Paper**

Directions:

1. Write your first line using 5 syllables. A haiku is a poem, but it doesn't rhyme, so don't worry about that.

2. Write your second line using 7 syllables.

3. Write your third line using 5 syllables again. This is the end of your poem, so make it count!

Fashion Statements

During the Edo period there was plenty of food and no fighting which resulted in many people's success and comfort. As a result the textile industry is one of the industries that expanded due to an increasing demand from the newly successful people and merchants.

Many of the fashions were specific to different groups, allowing someone to identify a person's role in society by simply looking at their clothing. The kimono, in particular, was used to identify one's social class. Military leaders would sport landscape designs on their kimonos, regular samurai would wear simpler shapes embroidered mostly around the waist, and women in the merchant class would wear family symbols on their shoulders with intricate designs below the midsection.

ADVENTURE 5

The kasa (笠) or hat was another way people used to signal who they were. The **amigasa** was a straw hat used for folk dances, monks would wear large woven hats in the shape of mushrooms, and samurai might wear **jingasa**, special hats made from a variety of materials like iron, copper or even wood. In many cases, these hats would have an emblem indicating the family or business the wearer wanted to represent.

Try to use your closet and put together an outfit that tells people who you are!

Time For Tea

Tea reached Japan from China in the 8th century as a medicinal beverage, but by the Edo period it had become a popular beverage for all people. **Sen no Rikyū** was a renowned tea master who heavily influenced the "**way of tea**." This ceremonial tea emphasized simplicity, directness, and honesty between both the host and their guest. By putting thought into every aspect of the beverage and its sharing, he was able to elevate a simple break for a drink into something deeply meaningful. To this day, tea ceremonies continue to be very common in Japan and act as a way for people to find calmness in our busy world.

While learning the methods of a tea ceremony takes a lot of time, you can apply the same ideas.

Directions:

1. Set up a time with someone you care about to share a drink or food.

2. Make sure that all screens are put away or silenced.

3. To make it special, use a tablecloth you don't normally use.

4. Make food and a tea that you don't have every day. For an extra special touch, put extra work into the presentation of the food.

Dr B.'s Note

By using care and attention, you can make a special experience for everyone involved.

Neat-o Edo Art

During the rule of the Tokugawa shogunate, a thriving urban culture led to a greater demand for the arts. The Edo period produced many painters that specialized in pottery, paintings, lacquerware (decorative paint coatings), calligraphy (beautiful writing), and even textiles! Many of the styles used during the Edo period are still popular in Japan today.

One popular genre of painting that emerged during the Edo period was called **ukiyo-e***. Ukiyo-e style paintings illustrated the everyday activities of the common people.

Many viewers of this style of art were everyday, or "middle-class," artisans and merchants. When people looked at these paintings, they often felt a shared experience with the artist.

*Say it like this:

ukiyo-e - "**you-key-oh-ey**"

Note: Syllables in Japanese words are generally spoken with equal stress, and for the same length of time.

Many of the ukiyo-e illustrations were of people, plants, or animals. Capturing something in a natural and realistic form was what made ukiyo-e art so relatable to observers.

47

Neat-o Edo Art

Materials:

- Newspaper
- Thick paper or canvas
- A cup
- Water
- Paints
- Paintbrushes
- A few paper towel sheets
- Pencil (optional)

 **Show off your artistic skills!**

Have your grown up take a photo, and share on social media using the hashtag:

#KnowYourAdventure

 KnowYourselfOAK   KnowYourselfOAK

Directions:

1. Try creating your own neat-o Edo art by using the home inventory materials. A ukiyo-e artwork in your home may look like a few of your family members having tea together. You could also take your art to an outdoors scene. Birds and trees make for great ukiyo-e subjects!

2. Once you have found your theme, set your workstation by laying down a few sheets of newspaper.

    Then place your canvas or thick paper on top.

3. Optional: It might be helpful to lightly outline your painting in pencil before beginning.

4. Prepare to paint by placing a cup of water, your paints, paintbrushes, and paper towels within an arm's reach.

5. Now you are ready! When switching paint colors, be sure to clean your paintbrush by dipping it in water and patting it dry with a paper towel.

Natural Talent

Japanese artists and poets drew inspiration from the world around them. By studying nature, they were able to capture beauty and wonder with the strokes of brush and pen. They weren't the only ones to do this – almost every human society has found inspiration in nature. What insights can you find?

Try spending a few minutes watching the environment, either in a park, your backyard, or even inside your house. Look for how plants move in the wind, how animals walk, and how things react to the light from the sun.

What things can you discover by simply sitting and watching?

_____
_____
_____
_____
_____
_____
_____
_____
_____
_____
_____
_____
_____

Fold Your Words

After you finish, check the answer key on page 118.

Across:

4. Japanese clothing that could be decorated to show the wearer's class.

5. Japan's capital has long been famous for these markets.

8. Shogunate from the early 1600s to the 1800s.

9. Both paintings and poetry in the 1600s to the 1800s tended to focus on this.

Down:

1. Welcome in Japanese.

2. Ukiyo-e is a genre of this.

3. A period of great cultural growth in Japan.

6. Popular Japanese Poem.

7. First Name of Famous Poet who used Zen Philosophy.

8. Largest city in Japan.

The Shogun Must Go On!

Good work, Adventurers!

Now that you have read some things about the history of the Edo Period in Japan, let's review what you have learned!

Try to fill in the blanks.

The Tokugawa shogunate ruled Japan from the early 1600s to the late __ __ __ __ __. During this time, the kingdom was divided into __ __ __ __ __ __ classes, made up of lords, samurai, farmers, __ __ __ __ __ __ __ __ , and traders.

During the Edo period, there was both cultural and __ __ __ __ __ __ __ growth in Japan. In modern-day Japan, we still see __ __ __ , food, and traditions that became popular in the Edo period. Edo is actually the original name of the largest modern city in the world, __ __ __ __ __ . During the Edo period Tokyo was a source of strategic shipping power for the __ __ __ __ __ __ __ shogunate.

Some of the cultural changes in Japan were related to _ _ _ _ _ _ , fashion, and art. Poetry took the form of a haiku (or "_ _ _ _ _" as it was known at the time). Fashion spread from samurai and those at the top of the social order to newly rich merchants. Many of the fashions matched to a person's specific role in Japanese society. In addition to the growing demand for _ _ _ _ _ _ _ , there was also an increased desire for the arts. Paintings, lacquers, pottery, and calligraphy were a few forms. One genre of painting was called "ukiyo-e", the style depicted the everyday life of the _ _ _ _ _ _ _ _ _ _ _ _ and could take the form of people, plants, landscapes, or _ _ _ _ _ _ _ (tigers were very popular).

You can check your answers using the keys on page 119.

THE DIGESTIVE SYSTEM

Know Your Digestive System

Gut Reaction

Did you know the digestive process starts before you even bite into your food?

You might have noticed that when you see or smell something delicious, your mouth starts "watering." That's **saliva**,* already preparing your mouth for **mastication*** (otherwise known as chewing). Your brain anticipates that you'll be eating soon, so it gets the digestion process started.

Your body actually uses two types of digestion: **Mechanical** and **Chemical** Digestion.

Mechanical digestion is when you break down food into smaller pieces, mainly by chewing. Such as when you bite off a piece of sushi. Chemical digestion, also starts with your mouth, but happens with the substances of your body further break down those foods for your body, mainly by the stomach.

*Say them like this:

saliva - "suh-LIVE-uh"
mastication - "mas-tih-KAY-shun"

The strongest syllable is always shown in **CAPITALS** and **red**.

As your teeth crush and grind this food, enzymes in your saliva begin breaking down the starches into sugar. Your teeth and saliva work together to create a small group of chewed food into a **bolus**.*

After you swallow the bolus, it passes through your esophagus, which has muscles that squeeze food down into your stomach through a series of contractions called **peristalsis**.*

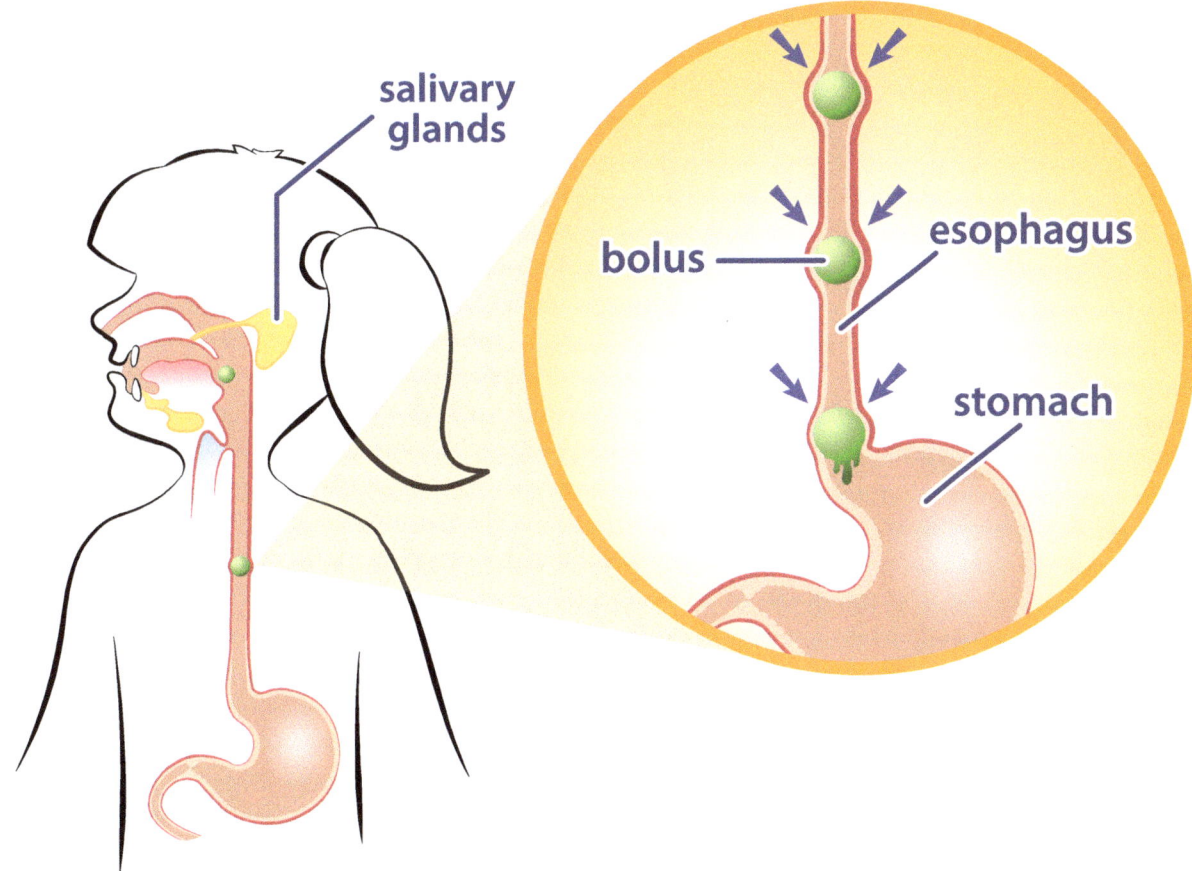

*Say them like this:

bolus - "BOW-luhs"
peristalsis - "pear-ih-STALL-sis"

The strongest syllable is always shown in CAPITALS and red.

Know Your Digestive System

Once in your stomach, the bolus mixes with gastric juices from your stomach and intestines to becomes a part-liquid mass called **chyme**,* made of semi-digested food, water, hydrochloric acid, and digestive enzymes.

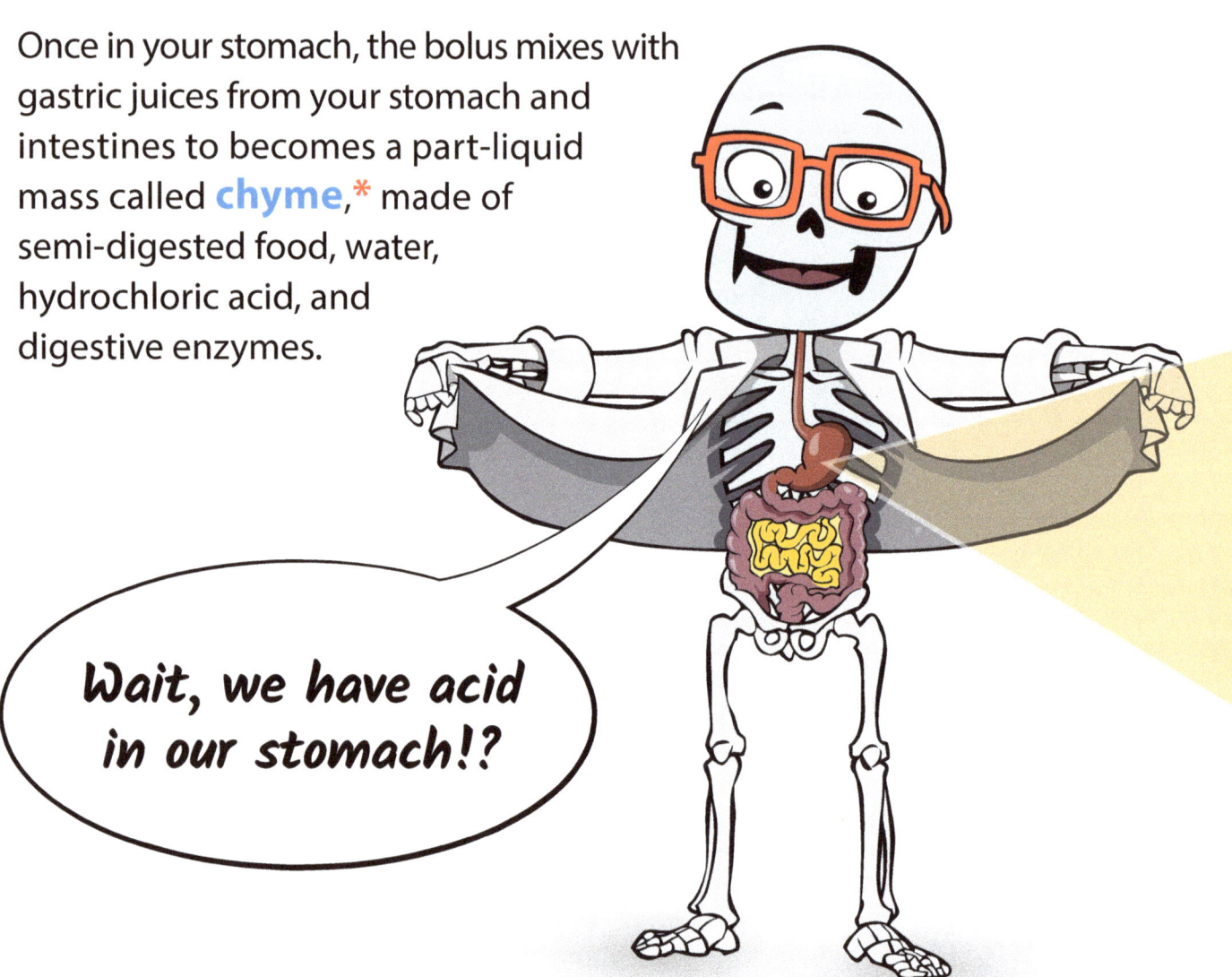

Wait, we have acid in our stomach!?

> *Say it like this:
>
> **chyme** - "**Ki-me**"
>
> The strongest syllable is always shown in **CAPITALS** and **red**.

Yes, stomach acid helps break down food as well as kill off any disease-causing bacteria that might enter the body when you eat.

However, the acid doesn't hurt you thanks to the protective **mucous membrane** in your stomach lining.

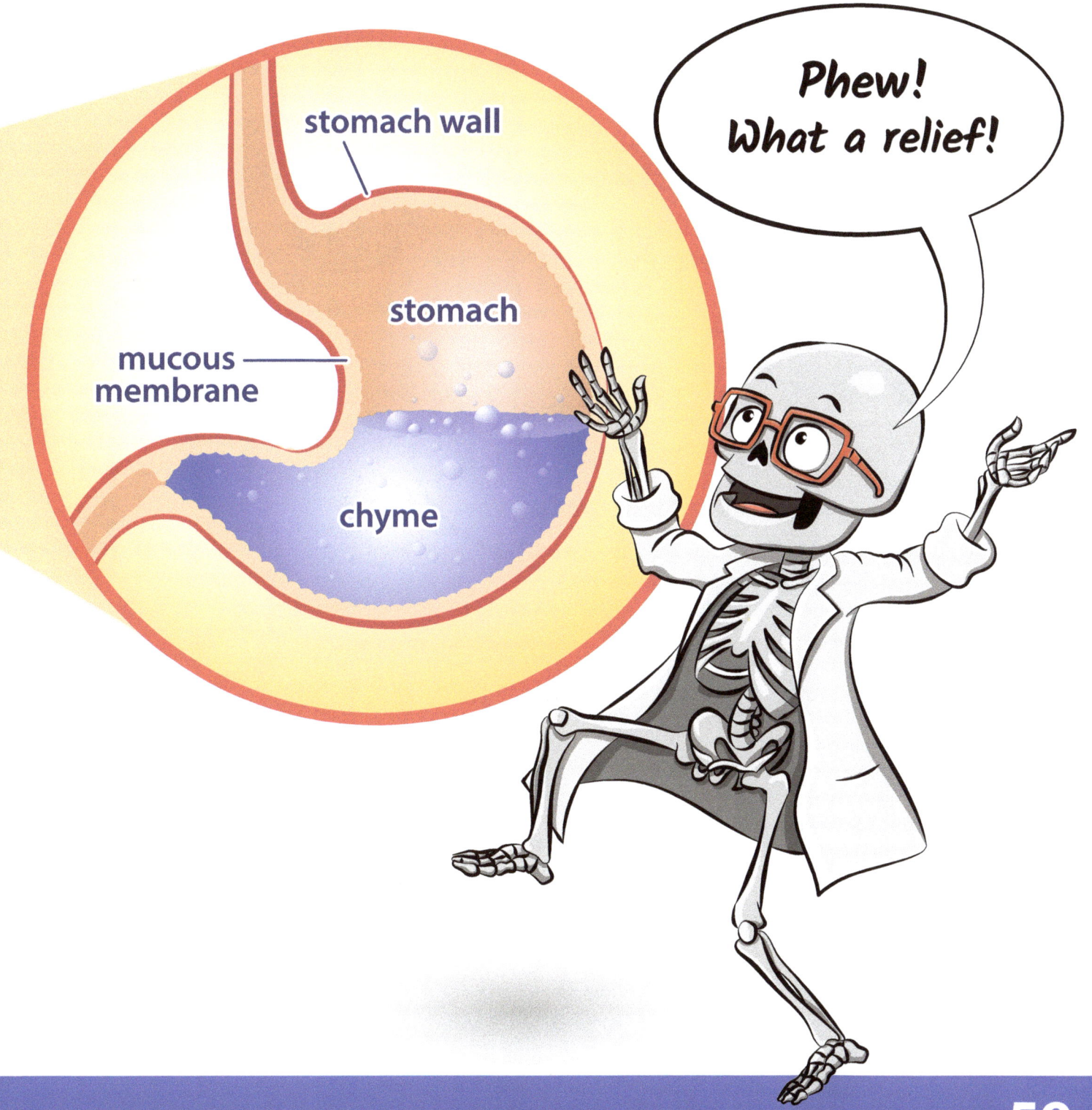

Know Your Digestive System

Afterwards, chyme passes into your small intestine where secretions from your liver and pancreas continue the chemical breakdown, eventually passing through your large intestine. Everything your body doesn't need is expelled as waste.

All these organs (your esophagus, stomach, and intestines) work together to make up the **alimentary*canal**, the long tube extending from your mouth to the end of your rectum.

The alimentary canal takes every meal and snack from eat to excrete!

*Say it like this:

alimentary - "al-ih-MEN-tuh-ree"

The strongest syllable is always shown in **CAPITALS** and **red**.

Fun Fact: The entire length of your digestive system is 20-30 feet!

The Alimentary Canal

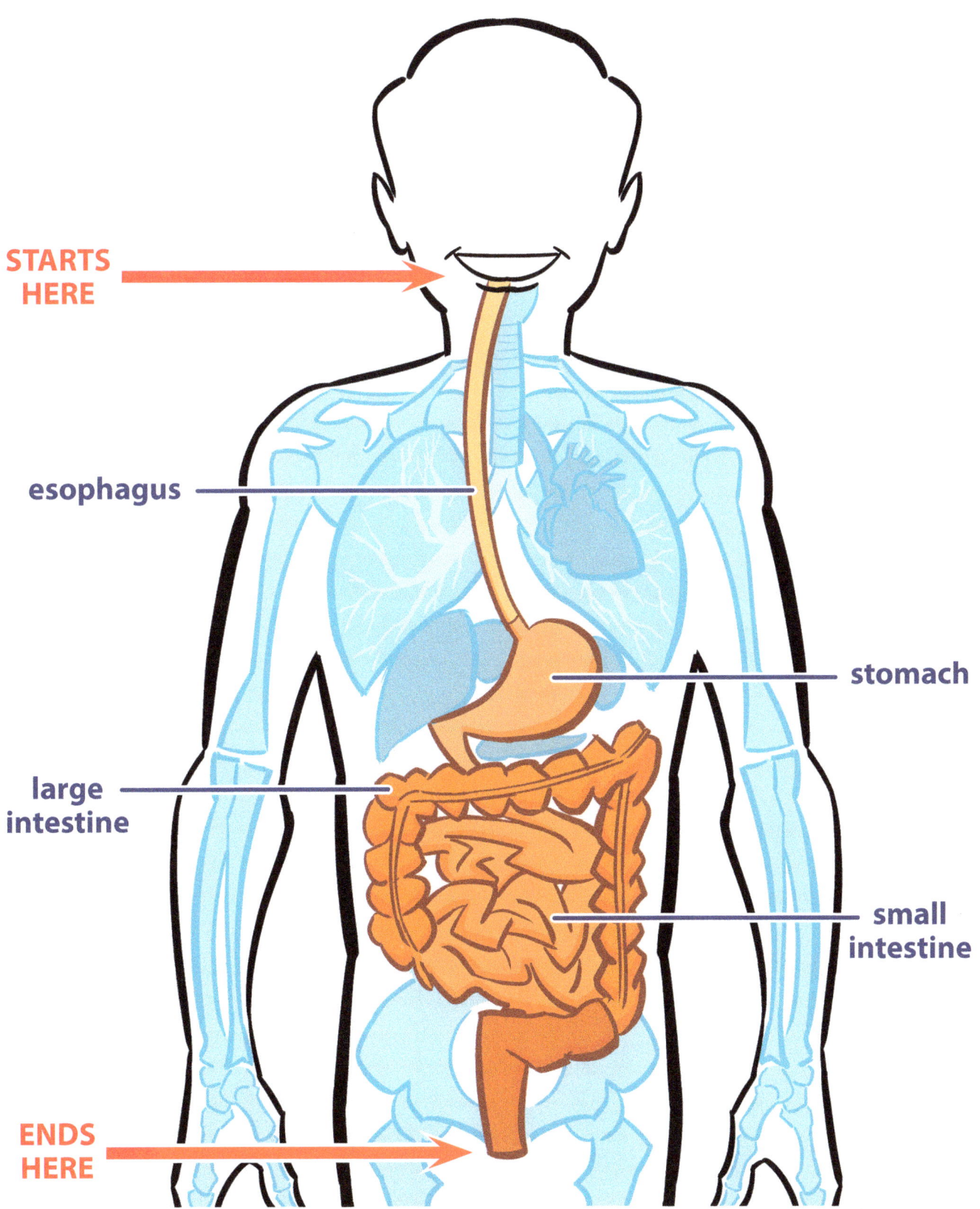

THE DIGESTIVE SYSTEM

Royal Flush

Your gut gets rid of waste as poop (known to scientists as **feces***).

It might be surprising to learn that poop is mostly water. The rest is an equal mix of bacteria (both living and dead), un-digestible vegetable fiber, and waste from other bodily processes.

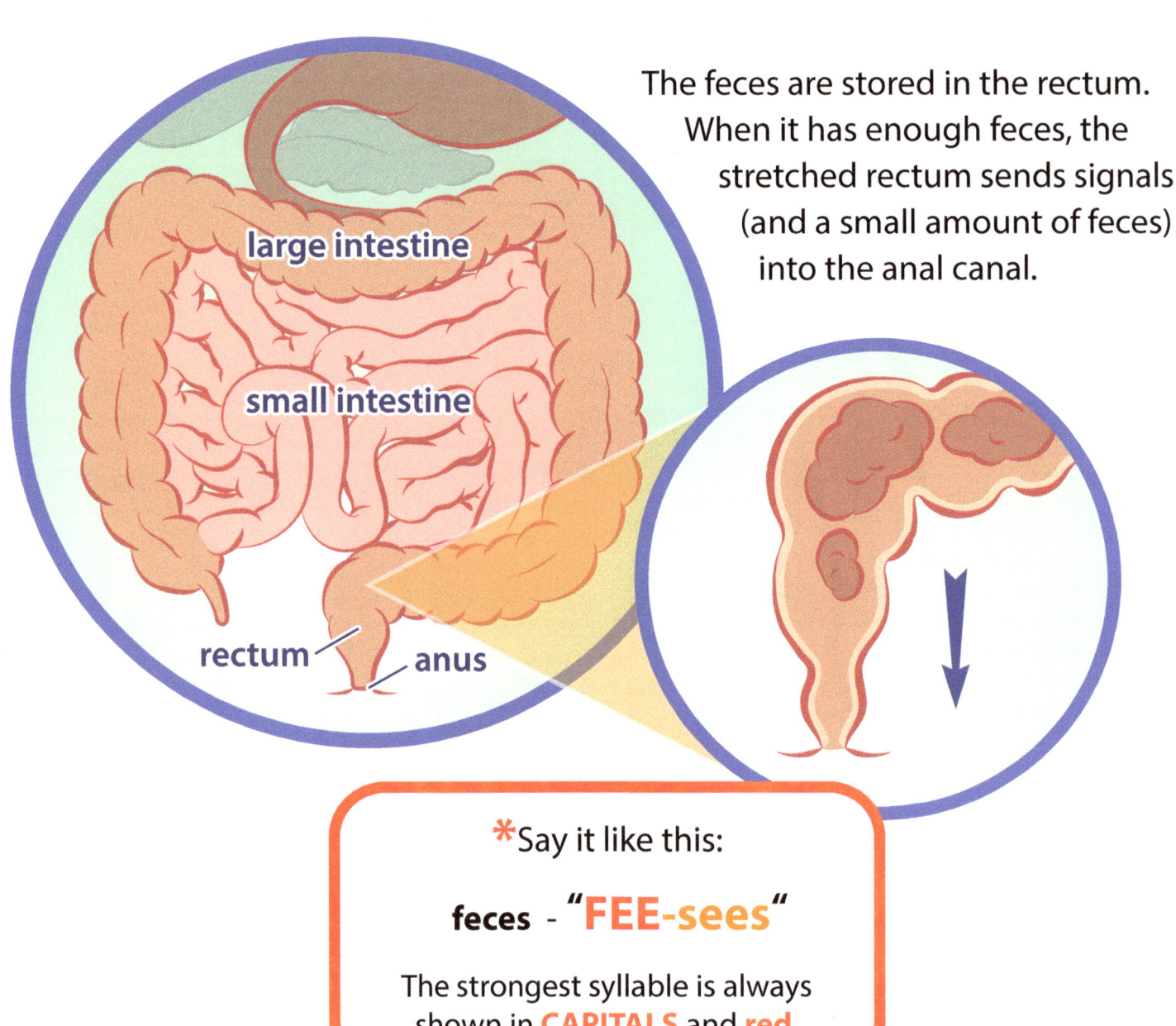

The feces are stored in the rectum. When it has enough feces, the stretched rectum sends signals (and a small amount of feces) into the anal canal.

*Say it like this:

feces - "**FEE-sees**"

The strongest syllable is always shown in **CAPITALS** and **red**.

62 ADVENTURE 5

The final sphincter is voluntary, so you can hold it until you get to a bathroom. Just make sure you don't hold it too long!

THE DIGESTIVE SYSTEM

Gas: A Work O' Fart

Gas is a byproduct of digestion.

Your body has to expel these gasses somehow. They leave the stomach in the form of burps, or from the intestines, in the form of **flatus***, or "farts".

A lot of this gas - the smelliest parts - comes from gases your gut bacteria make while they're digesting things in your colon.

Foods with lots of fiber tend to produce more gas.

Foods like beans, cabbage, cheese, and eggs contain a lot of sulfur, a smelly chemical that makes flatus smell worse.

> *Say it like this:
>
> flatus - "FLAY-tuss"
>
> The strongest syllable is always shown in CAPITALS and red.

This Adventure's Touchstone: where anatomy, physiology, and psychology all come together.

Life Savor

Have you ever heard of "mindful eating?"

Wait a minute—what does your mind have to do with cheese pizza or carrots or a cookie? Let's **ruminate*** on that for a bit.

Eating a meal might take a half hour or so, but digestion takes even longer. That pizza's journey through your stomach and intestinal tract lasts 6–8 hours! Food has to be broken down mechanically (into little bits) and then chemically (into even smaller bits) so your body can use it for energy.

*Say it like this:

ruminate - "ROOM-ih-nate"

The strongest syllable is always shown in CAPITALS and red.

ADVENTURE 5

But hasty ingestion doesn't make this process go by more quickly or easily. In fact, it's slowing down and paying attention that can really help.

Essentially, mindful eating means being aware of what you're eating and why. Our bodies change food into fuel, but from the first bite to the last, food can change both our bodies and our minds.

Sometimes, we say things like,

"**That book was a lot to digest**"
or
"**I'm still digesting the big news**".

What would "**digest**" mean here?

Can you think of some other sayings that relate to food or eating?

THE DIGESTIVE SYSTEM

67

Kaiseki

Kaiseki* is a traditional Japanese dinner made of multiple courses, including, for example, appetizers, soup, sashimi (raw fish), and a rice dish. The food is created and prepared with great care and attention, and eating it is meant to be an enjoyable and memorable experience.

"The idea and philosophy behind kaiseki is the appreciation of nature, seasonality, and focus into the moment that is in front of you, the moment at hand..."

Chef Niki Nakayama

> ***Say it like this:**
>
> **kaiseki** - **"kie-sek-ee"**
>
> Note: Syllables in Japanese words are generally spoken with equal stress, and for the same length of time.

Here's some food for thought:

1. **Tune into your body's signals.**

   Are you hungry, or used to snacking in front of the TV? Did you wait too long to eat—are you irritable, is your stomach rumbling?

2. **Use all your senses at mealtime:**

   Smell, sight, touch, taste, even sound!
   Notice the colors, the smoothness, or crunch.
   How does a tomato smell fresh off the vine, or as it cooks?
   And of course, how does it taste?

3. **Take your time.**

   Your brain needs about 20 minutes to realize you're full. Scarfing down lunch means you might think your stomach is still empty when it isn't.

4. **Feed more than your belly.**

   We eat for nutrition, but also celebration, and tradition. How can a meal for a birthday, holiday, or other special occasion be eating mindfully?

THE DIGESTIVE SYSTEM

Digestion Detective

Take a closer look at your digestive system! The yarn or string represents the length of your real-life intestines. Ready, detective?

Materials:

- **Posterboard**
- **Markers, craft paint, or other materials**
- **2 pieces of yarn or string, 6 meters and 3 meters long**
- **Liquid glue**

Directions:

1. Lay your poster board on the floor. Then have a friend help trace the upper half of your body on the board using a marker.

2. Now comes the fun! Fill in the poster version of you with digestive system body parts (esophagus, stomach, large intestine, and small intestine). Use the diagram from Know Your Digestive System to help you with the placement.

3. Paint your esophagus leading away from your mouth and into the stomach.

4. Then, illustrate your stomach by using paint or other craft materials.

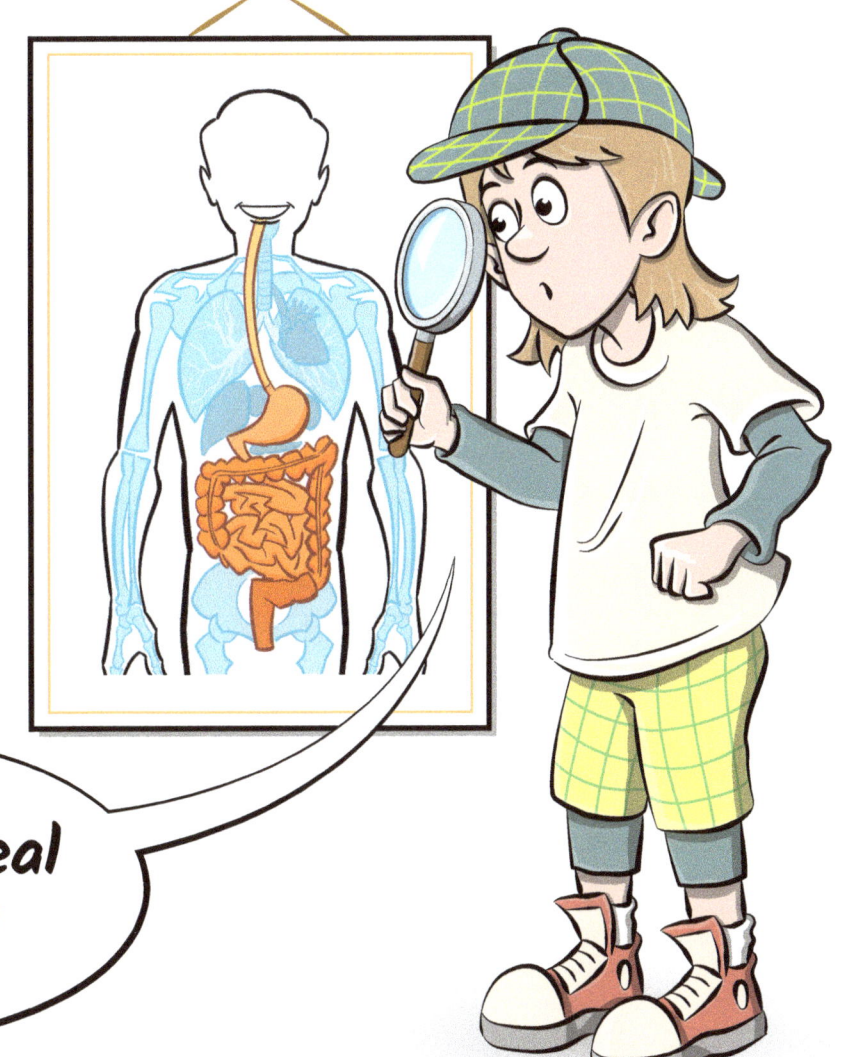

This is a real mystery!

THE DIGESTIVE SYSTEM

71

Digestion Detective

5. Connect the stomach to the small intestine by adding the 3-meter long yarn to the board. Hold it in place using glue.

6. Finally, join in the large intestine (your 6-meter long piece of yarn). The large intestine should circle around the small intestine.
Apply glue to secure the yarn.

7. Make your digestion diagram uniquely you by dressing it up with more paint or other craft materials.

8. Stand back and admire your digestive system diagram.

What are a couple of things you detect after observing your life-size digestion diagram? Jot down your thoughts and observations.

_____

_____

_____

_____

_____

_____

**Did you know that
the small intestine has a diameter of
3.5 to 4.5 centimeters
and
the large intestine has a diameter of
4 to 6 centimeters?**

Can you find any items with a circular opening in your house that measure this size in diameter?

Check around your house for these items (think bottles, jars, or socks!).

Follow Your Food

Now that you have a pretty decent idea of what your digestive system looks like, let's further explore how food moves through the body.

Materials:

- 1 small resealable plastic sandwich bag
- Cereal
- Water
- Soda
- 1 nylon stocking
- Scissors

Follow your food from ingestion to excretion!

ADVENTURE 5

Directions:

1. Seal a small handful of cereal in the plastic sandwich bag, along with a spoonful of water. The water acts like saliva, which contains an enzyme called **amylase*** that helps break up carbohydrates.

   The action of this enzyme is called **chemical digestion**.

2. Mash up the mixture by gently squeezing the bag. This is similar to what occurs when you chew up food with your mouth and teeth.

   This physical breakdown of food is called **mechanical digestion**.

> *Say it like this:
>
> **amylase** - "AM-ill-ayz"
>
> The strongest syllable is always shown in CAPITALS and red.

Follow Your Food

3. Once the food is ground up by your mouth and teeth, it travels down your esophagus and into your stomach. Open the bag and add a half cup of soda to the mix to represent this step. The soda acts like stomach acid, which further breaks down nutrients.

4. From the stomach, the food (now called 'chyme') travels to the small intestine.

   In the small intestine, your body absorbs nutrients from the broken down food before making its way to the large intestine.

   In the large intestine, water and bile salts are removed from the chyme. Finally, food exits the body.

   Follow your food by grabbing a pair of scissors and cutting off the bottom foot of the nylon stocking. Place your homemade chyme into the top of the nylons (acting as the intestines). Move the chyme through the stocking and out the bottom hole (you may want to do this over a sink!).

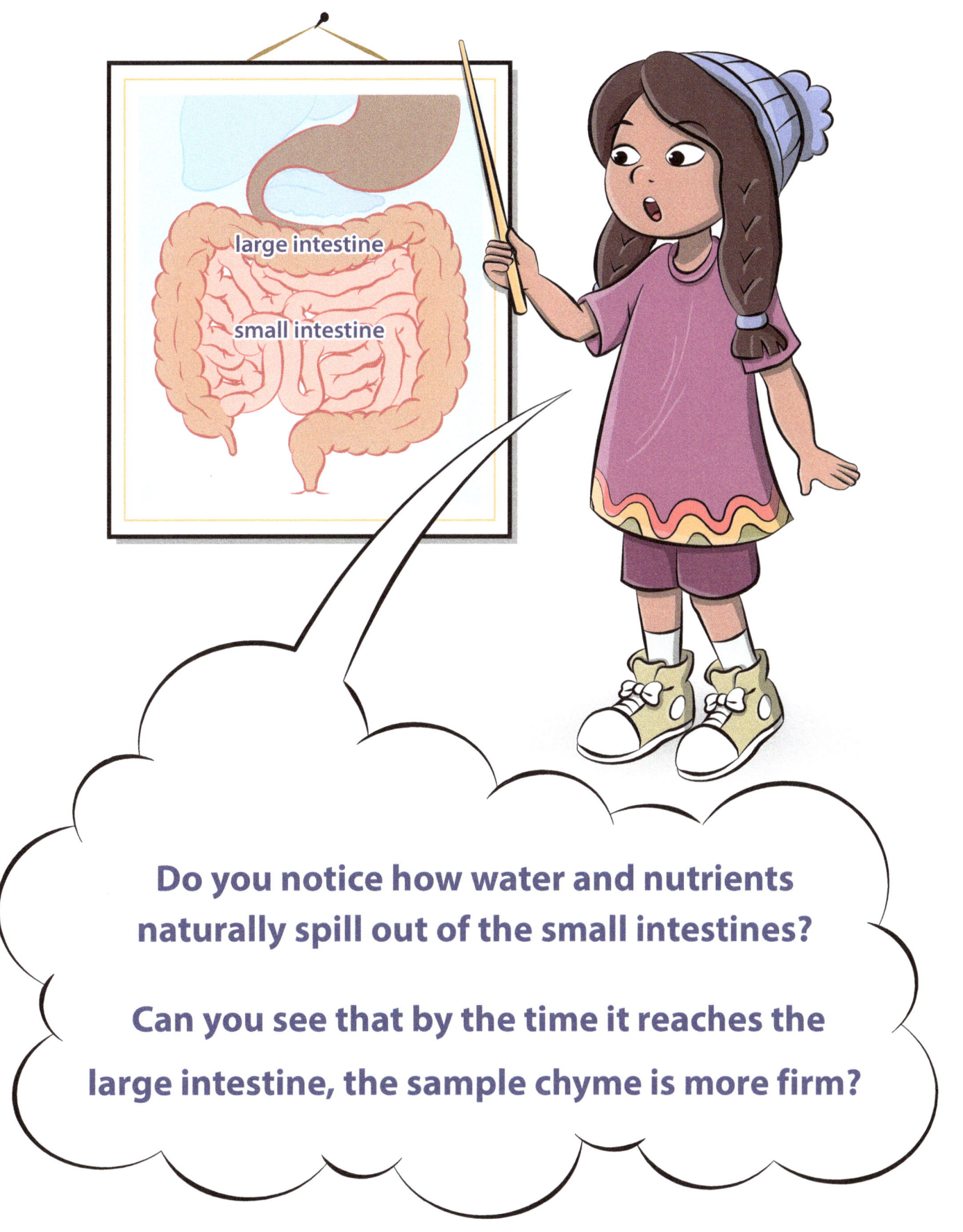

Do you notice how water and nutrients naturally spill out of the small intestines?

Can you see that by the time it reaches the large intestine, the sample chyme is more firm?

A Fuzzy Situation

What does the inside of your small intestine look like?

Take a look inside of this fuzzy situation using a sock and a few home-inventory materials. The fuzzy sock represent the villi and microvilli that help to transport nutrients from the small intestines and into the rest of your body.

Materials:

- **A fuzzy sock** (that you don't need. Check your lost sock collection!)
- **Scissors**
- **Square-inch grid paper**
- **A pencil or pen**

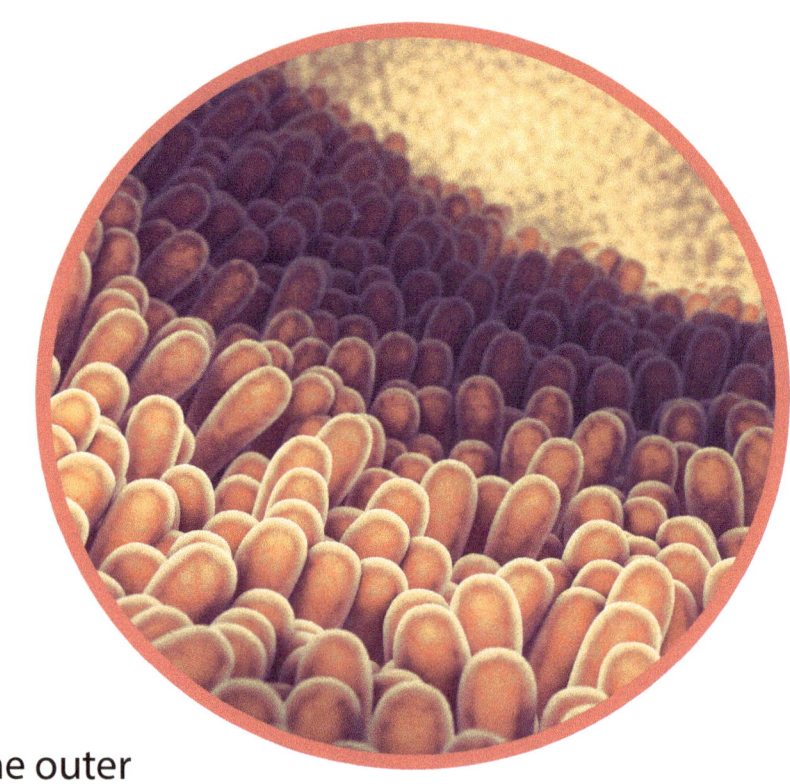

Directions:

1. Turn your fuzzy sock inside out. The outer part of the sock represents the smooth texture of the small intestine. Look inside the opening. The "fuzzies" represent the small, finger-like folds called **villi***, and smaller hair-like projections called **microvilli***, that line the inside of the small intestine. Cut the sock open lengthwise and lay it down on top of a piece of square-inch grid paper. Trace around the material using a pencil or pen and count the total number of square inches it covers. For partial square inches, imagine putting them together to make a whole square inch for an approximate area measure.

> *Say them like this:
>
> villi - "VIH-lie"
>
> microvilli - "MY-crow-VIH-lie"
>
> The strongest syllable is always shown in CAPITALS and red.

2. The villi and microvilli increase the surface area of the interior of the small intestine to almost the size of half of a basketball court! This helps to absorb a large amount of nutrients in a short amount of time (about 3 to 6 hours).
   That's 338,400 square inches!

   How many fuzzy socks would you have to cut open and lay down to cover half a basketball court? Divide 338,400 by your sock area to find out.

3. Now imagine that number of socks with villi and microvilli working to absorb nutrients from the food you eat!

Dr B.'s Note

Your small intestine sends nearly everything it absorbs directly to the liver through a special system of blood vessels, allowing the liver cells to break down toxins (such as poisons put in your rice by your enemies!).

THE DIGESTIVE SYSTEM

Numbers and Nutrients

Do you know why the number 7 is so lucky?

It's probably not just a myth. It's probability! Probability is a calculation of how likely it is (the "odds") that something is going to happen.

Understanding probability is important for a huge number of scientific fields from astronomy to biology. When you eat, your mouth might be excited for the different tastes but your body is looking for the right nutrients. Proteins, fats, carbohydrates, vitamins, minerals and water are all essential to keeping your body working right, and the digestive system is there to break down the food you eat into parts your body can use to keep itself running.

Eating a healthy diet is important to make sure you get everything your body requires, and your digestive system is very good and taking everything it can use out of your food.

It isn't always fast, however. Food can take from two to five days to pass through your body. How long it takes depends on what you ate, how your body in particular works, and it's also a little bit random.

ADVENTURE 5

When you roll two dice, the most common number to get as the total of both dice is seven, but that doesn't mean the other numbers won't show up also.

A lot of biology has processes that work within a specific range like that, where some things are more common, but just because you are a bit off the standard doesn't necessarily mean anything is wrong.

Try playing this game to get a sense of how probability works.

Numbers and Nutrients

Materials:

- Two 6-sided dice
- Paper
- Pen
- Something to use as a token, like beads or pennies
- A second player

	Die #1					
Die #2	1	2	3	4	5	6
1						
2		4				
3		5				
4			7			
5						
6						

Directions:

1. Make a chart like the one in the previous page using your paper. Make sure to leave enough space for each box to fit what you are using for a token!

2. Fill in the boxes by adding the numbers at the top and the left of each column and row. We got you started with three of them.

3. The younger player picks either below or above seven to be their side of the board.

4. The younger player rolls the dice to start the game.

    a. If they roll a combination that is on their side of the board, put a token on it.

    b. If they roll a combination on the other player's side, the other player gets the dice.

    c. If they roll a 7, they can remove a token from their opponent's side of the board!

5. Continue until one player has ten tokens placed.

That player is the winner!

Whole-y Guacamole!

Nutrients are essential for your organs to function.

Your digestive system helps begin the process of delivering important nutrients to every part of your body. Make sure you get enough nutrient-dense foods by aiming to have a serving of whole grains, fruits, and/ or vegetables at every meal.

Your organs have lots of tasks to complete inside the body! When you fuel yourself with healthy foods, you help your body do what it needs to do, plus have energy for playing games and having fun!

As the great Greek philosopher Hippocrates once said,

"Let food be thy medicine, and medicine be thy food."

Do a search and see if you can point to any "whole" foods in your home or during your next trip to the grocery store. Jot down a list of any "whole" foods you see in your home or in the grocery store.

"**Whole**" foods are unprocessed foods that come straight from a plant source. A few great (and tasty) examples of these are beans, rice, carrots, corn on the cob, celery, broccoli, apricots, oranges, pineapple, and avocado (yum - guacamole!).

What are some of your favorite 'whole' foods?

_____

_____

_____

_____

Do you notice a difference in how you feel when you eat them compared to when you eat junk food? Write down your thoughts.

_____

_____

_____

_____

Digestive Dash

All the components of your digestive system have gotten jumbled up! Find them to get things flowing.

ACID

ENZYMES

INTESTINES

SALIVA

CHYME

ESOPHAGUS

MASTICATION

STOMACH

DIGESTION

EXCRETION

MICROVILLI

VILLI

Answer keys on page 120.

```
I V Y N L E O U E N X Q L N I
E D T M Z M V J O Z S C W K K
N A U A O E F I C E J X J I Z
Z N L S T A T I N C H Y M E V
Y M W T V E U I W F Y G A C J
M C S I R W T D E H G O L E D
E O L C Y S S L M V Y P U G I
S A X A E H O T C Q I Q K H G
S E S T M I C R O V I L L I E
A Z N I G C X I O M O N L X S
C I J O Z W J W M A A E A I T
I E P N T C K S Q O Y C P I I
D B H Q W X K M W K V W H O
G Y E S O P H A G U S S A U N
B P P P K O N T N Q F H I C C R
```

THE DIGESTIVE SYSTEM — 87

Digest Your Knowledge

Now that you have learned all about the digestive system, let's put your knowledge to the test!

Fill in the blanks below.

When your mouth waters, that's your body creating __ __ __ __ __ which helps for chewing or __ __ __ __ __ __ __ __ __ __ __ __. This is the first step of your __ __ __ __ __ __ __ __ __ process. There are two types of digestion that your body uses, __ __ __ __ __ __ __ __ __ and __ __ __ __ __ __ __ __.

Your mouth uses both to convert food into a __ __ __ __ __ to swallow. Afterwards, it passes through your esophagus in a process called __ __ __ __ __ __ __ __ __ __ __.

In your stomach, the bolus mixes with __ __ __ __ __ __ __ __ __ __ __ __ __ __ and becomes __ __ __ __ __. The stomach __ __ __ __ helps break down food and kill bacteria. Don't worry though, your __ __ __ __ __ __ membrane keeps you safe.

Check the answer keys on page 121!

coloring opportunity

THE DIGESTIVE SYSTEM

Know Your Appetite

Experience Japanese Foods

Japan's food culture is rice-centered, and this starchy base is generally combined with pork, tofu, or seafood. Legend has it that sushi was created when an elderly woman began hiding her pots of rice in birds' nests to protect them from thieves. After a time, she collected her pots and discovered that the rice, because it had been left out for so long, had begun to ferment.

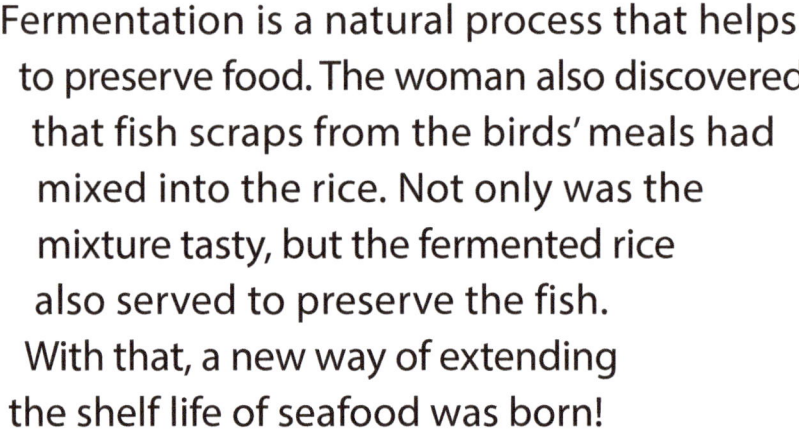

Fermentation is a natural process that helps to preserve food. The woman also discovered that fish scraps from the birds' meals had mixed into the rice. Not only was the mixture tasty, but the fermented rice also served to preserve the fish. With that, a new way of extending the shelf life of seafood was born!

Pinky's Hint:

Read through the entire recipe before beginning to prepare food. This way, you'll know what equipment and ingredients are needed, and you'll be familiar with the steps involved.

 Whenever you see the chef's hat icon, it means **you'll need an adult's help**.

(Itadakimasu)*

That means **"Thanks for the food!"** in Japanese.

> *Say it like this:
> **"ee-tah-dah-key-mahs"**
>
> Note: Syllables in Japanese words are generally spoken with equal stress, and for the same length of time.

Recipes and food knowledge provided by Chef Polly Legendre of La Gourmande Catering.

Pickled Cucumbers

There weren't any refrigerators in Edo Period Japan, so pickling (helped by fermentation) became a popular way to help vegetables last longer and, in some cases, make them healthier. Some pickles help with digestion, while others contain important vitamins. Pickled rice bran (the hull around the grain) has important B vitamins that were scarce in the Japanese diet.

Here's a recipe to help get you familiar with a basic but delicious pickling technique.

 **Prep time:** 10 minutes

 **Wait time:** 10 minutes

 **Makes about 1 ½ cups of pickled cucumbers**

Ingredients:

- 2 or 3 small Japanese or Persian cucumbers (if you can't find the small ones, 1 regular size English cucumber will work as well)
- 2 tablespoons salt
- ¼ cup rice vinegar
- 2 teaspoons sugar
- 1 pinch of salt
- 1 tablespoon sesame seeds

Preparation:

1. Wash the cucumbers. Remove the stems ends, then slice into thin coins. Place the slices into a bowl.

2. Sprinkle the 2 tablespoons of salt over the cucumbers and let them sit for 5 to 10 minutes.

3. Rinse off the salt with cold water, and drain the cucumbers.

4. Combine the rice vinegar, sugar, and pinch of salt in a bowl.

5. Gently mix the cucumbers in with the vinegar mix.

6. The pickled cucumbers are ready to eat now, but they can be kept in the refrigerator for up to 4 days. The pickles will become a bit rubbery and less tasty the longer they are kept in the refrigerator.

Sushi Rolls

Vegetable Maki

The Japanese word "sushi" refers to the combination of rice and vinegar. "Sashimi" refers to the thinly sliced raw fish found in sushi rolls. Maki sushi—or norimaki—is rice, vegetables, and/or fish rolled in seaweed.

You will need plastic wrap and your bamboo mat.

Don't have a bamboo mat? Use the towel trick! A thick dish towel acts just like a bamboo mat. Place your plastic wrap piece on top of the towel. Put the seaweed and rice down on the plastic and make into a beautiful roll.

Prep including rice cooking: 40 minutes

Rolling time: 30 minutes

Makes 4 servings

Prep the rice

Ingredients:

- 2 cups short-grain or sushi rice (only short- or medium-grain will work)
- 2 1/4 cups water (plus what is needed to rinse the rice before cooking)
- 2 tablespoons rice vinegar
- 2 tablespoons sugar
- 1 teaspoon salt
- 4 Nori (seaweed) sheets for sushi rolls
- 2–3 cups vegetables such as sliced or shredded carrots, cucumbers, or radishes (you might have extra)

Roll that sushi!

Eat that sushi!

Sushi Rolls - Vegetable Maki

Preparation:

Making Sushi Rice

1. Place the rice in a mixing bowl and rinse with cold water. Repeat 2 or 3 times until the water that you pour out is clear.

2. Place the rice and a fresh 2 ¼ cups of water in a saucepan, then bring to a boil over high heat, uncovered. (If you have a rice cooker, follow the instructions that came with the cooker.)

3. Once the rice has started to boil, reduce the heat to the lowest setting and cover. Let it cook for 15 minutes.

4. While the rice is cooking, prepare the vinegar mixture by combining the rice vinegar, sugar, and salt in a small bowl.

5. Remove the saucepan with the rice from the heat, but let it stand, still covered, for another 10 minutes.

6. Once the rice has "rested," transfer it to a larger wooden or glass mixing bowl. Add the rice vinegar mixture and gently fold it into the rice. Let the rice cool completely.

Preparation:

Ready to Roll?

1. Lay down a sheet of plastic wrap on top of your bamboo mat. This will help keep it clean. Next, place a sheet of nori on your sushi mat, rough side up.

2. Wash your hands, then use them to spread the rice evenly onto the sheet, leaving about a 1/2 inch of nori empty along the upper edge.

3. Place thin strips (about the width of your pinky finger) of vegetable filling in a line across the center of the rice. If you use too much filling, your roll won't close!

4. Starting on the side closest to you, lift the lower edge of the bamboo mat and use it to fold the nori and rice over the filling. Make sure the rice sticks together and the filling stays in place.

Sushi Rolls - Vegetable Maki

5. Roll the sushi by tucking the front edge of the nori into the roll. Then, begin rolling the sushi while removing the mat. Roll slowly to ensure the sushi roll is coming out even.

6. Tighten the roll so that the filling doesn't fall out. Do this by rolling it back and forth in the mat to secure and seal it.

 **7.** Finally, cut the roll into 6 to 8 pieces using a sharp, wet knife. Your rolls are ready to serve!

 **Show off your cooking skills!**

Have your grown up take a photo, and share on social media using the hashtag:

#KnowYourAdventure

 KnowYourselfOAK   KnowYourselfOAK

ADVENTURE 5

Thoughts for Young Chefs

What did you learn about Japanese food that you didn't know before this Adventure?

_____

_____

_____

_____

_____

_____

_____

_____

_____

_____

THE DIGESTIVE SYSTEM

Thoughts for Young Chefs

When Japanese chefs first opened sushi restaurants in California, they started using avocados, an ingredient that wasn't available in Japan, in their sushi rolls.

Can you think of other dishes that use different ingredients in different parts of the world?

_____

_____

_____

_____

_____

_____

_____

_____

What type of food are you inspired to make next?

Food for Thought in Feudal Japan

Adventurer, great job working your way through this guide!

I tip my *kasa* to you! Before you close this adventure, can you answer just a few "food for thought" questions below?

In this adventure, we learned about different roles that existed in Japan's feudal system. **What kind of positions can you think of that exist in modern society?**

_____

_____

_____

_____

_____

_____

_____

What popular art, foods, or traditions do you think future societies will see as characteristic of our times? Write down your thoughts below!

Food for Thought in Feudal Japan
(continuation)

As you learned in the adventure, the digestive system is a complicated process. It probably isn't a surprise that it can sometimes have issues that cause stomach aches or indigestion. There are many different causes of this discomfort, you can get it from eating too much or eating food that your body in particular dislikes. Think of a time that you have felt like you had a stomach ache. **Where in the process do you think things went wrong?**

_____

_____

_____

_____

_____

_____

_____

_____

Do you think there is anything you could do in the future to avoid that problem?

_____
_____
_____
_____
_____
_____
_____
_____
_____
_____
_____

Further Reading

Nonfiction

Horrible Science: Disgusting Digestion
Arnold, Nick and Tony De Saulles; **Ages 8+**

The Quest to Digest
Corcoran, Mary and Jef Czekaj; **Ages 8+**

*You Wouldn't Want to Be a Samurai!:
A Deadly Career You'd Rather Not Pursue*
MacDonald, Fiona; **Ages 8+**

*The Lucky Escape: An Imaginative Journey
Through the Digestive System
(Human Body Detectives)*
Manley, Heather and Jessica Swift; **Ages 6+**

Samurai Rising: The Epic Life of Minamoto Yoshitsune
Turner, Pamela S. and Gareth Hinds; **Ages 10+**

*Digestive System Part 1: Crash Course Anatomy & Physiology #33
(followed by Parts 2 and 3)*
http://bit.ly/DigestiveSystem_CrashCourse; Ages 12+

Further Reading

Fiction

The Samurai's Tale
Haugaard, Erik C.; **Ages 10+**

The Ghost in the Tokaido Inn
Hoobler, Dorothy and Thomas; **Ages 10+**

The Samurai and the Long-Nosed Devils
Namioka, Lensey; **Ages 8+**

Usagi Yojimbo Saga
Sakai, Stan; **Ages 10+**

Yuki and the One Thousand Carriers
Whelan, Gloria; **Ages 6+**

GUT The Inside Story of Our Body's Most Underrated Organ
Enders, Giulia; **Ages 12+**

Gulp: Adventures on the Alimentary Canal
Roach, Mary; **Ages 12+**

Samurai: An Illustrated History
Kure, Mitsuo; **Ages 12+**

47 Ronin
Richardson, Mike and Stan Sakai; **Ages 13+**

Musashi: An Epic Novel of the Samurai Era
Yoshikawa, Eiji and Charles Terry; **Ages 12+**

ADVENTURE 5

THE DIGESTIVE SYSTEM

THE DIGESTIVE SYSTEM

NEXT
6 The Immune System

Get Ready to Visit Arthurian England

HEAR YE, HEAR YE, in this adventure, you'll arrive in fictional sixth-century England and the legendary lands of Camelot. Unearth the particulars about the immune system.

Get to Know...

Naz

Wry, funny, intuitive, and confident, Naz's sarcasm can come off as insulting to some, though she's really just not afraid to say what's on everyone's mind. A bibliophile, she believes that literacy leads to mindfulness and wants people around her to be well read. She is also quite interested in art history and is the only member of the Loops Crew that longboards.

Age: 12

Favorite Color: Purple

Enjoys: Reading, Learning about her culture, Looking at art, Spending time with her family

Admires:
Tori Amos (singer)
Sharice Davids (politician)
Frida Kahlo (painter)
Luis Fonsi (singer)

Favorite Quote:
No act of kindness, no matter how small, is ever wasted.
— *Aesop*

Answer Keys

Fold Your Words

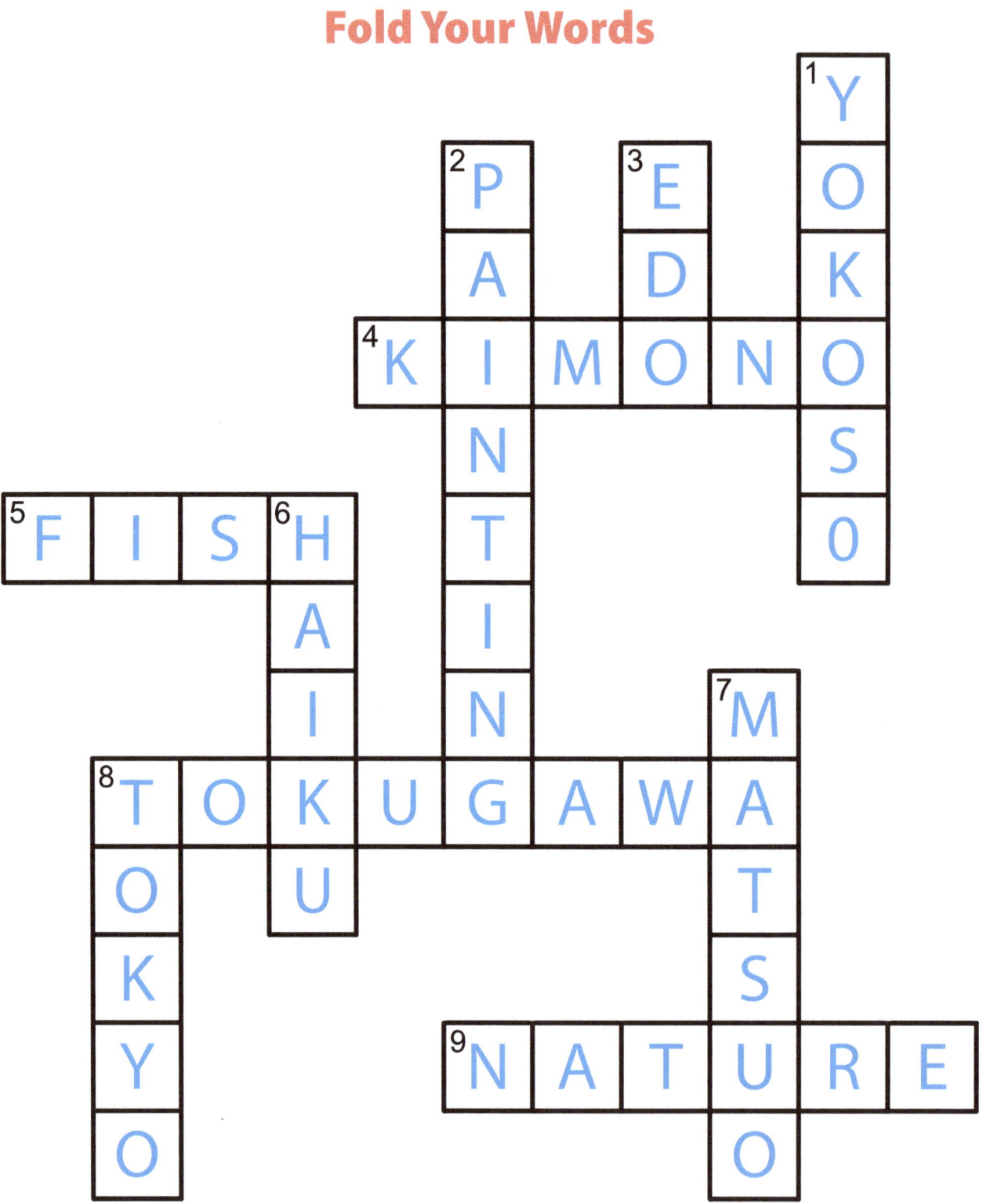

The Shogun Must Go On!

The Tokugawa shogunate ruled Japan from the early 1600s to the late 1800s. During this time, the kingdom was divided into social classes, made up of lords, samurai, farmers, artisans, and traders.

During the Edo period, there was both cultural and economic growth in Japan. In modern-day Japan, we still see art, food, and traditions that become popular in the Edo period. Edo is actually the original name of the largest modern city in the world, Tokyo. During the Edo period Tokyo was a source of strategic shipping power for the Tokugawa shogunate.

Some of the cultural changes in Japan were related to poetry, fashion, and art. Poetry took the form of a haiku (or "hokku" as it was known at the time). Fashion spread from samurai and those at the top of the social order to newly rich merchants. Many of the fashions matched to a person's specific role in Japanese society. In addition to the growing demand for textiles, there was also an increased desire for the arts. Paintings, lacquers, pottery, and calligraphy were a few forms. One genre of painting was called "ukiyo-e", the style depicted the everyday life of the common people and could take the form of people, plants, landscapes, or animals (tigers were very popular).

Answer Keys

Digestive Dash

I	V	Y	N	L	E	O	U	E	N	X	Q	L	N	I		
E	D	T	M	Z	M	V	J	O	Z	S	C	W	K	K		
N	A	U	A	O	E	F	I	C	E	J	X	J	I	Z		
Z	N	L	S	T	A	T	I	N	C	H	Y	M	E	V		
Y	M	W	T	V	E	U	I	W	F	Y	G	A	C	J		
M	C	S	I	R	W	T	D	E	H	G	O	L	E	D		
E	O	L	C	Y	S	S	L	M	V	Y	P	U	G	I		
S	A	X	A	E	H	O	T	C	Q	I	Q	K	H	G		
S	E	S	T	M	I	C	R	O	V	I	L	L	I	E		
A	Z	N	I	G	C	X	I	O	M	O	N	L	X	S		
C	I	J	O	Z	W	J	W	M	A	A	E	A	I	T		
I	E	P	N	T	C	K	S	Q	O	Y	C	P	I	I		
D	B	B	H	Q	W	X	K	M	K	V	K	W	H	O		
G	Y	E	S	O	P	H	A	G	U	S	S	A	U	N		
B	P	P	K	O	N	T	N	Q	F	H	I	C	C	R		

Digest Your Knowledge

When your mouth waters, that's your body creating saliva which helps for chewing or mastication. This is the first step of your digestion process. There are two types of digestion that your body uses, mechanical and chemical.

Your mouth uses both to convert food into a bolus to swallow. Afterwards, it passes through your esophagus in a process called peristalsis.

In your stomach, the bolus mixes with gastric juices and becomes chyme. The stomach acid helps break down food and kill bacteria. Don't worry though, your mucous membrane keeps you safe.

CREATED WITH LOVE

BY THE

KNOW YOURSELF TEAM

KnowYourself.com   KnowYourselfOAK   KnowYourselfOAK

www.ingramcontent.com/pod-product-compliance
Lightning Source LLC
LaVergne TN
LVHW070951070426
835507LV00031B/3494